Transforming Animals

TURNING INTO A NEWT

by Tyler Gieseke

Cody Koala
An Imprint of Pop!
popbooksonline.com

abdobooks.com
Published by Pop!, a division of ABDO, PO Box 398166, Minneapolis, Minnesota 55439. Copyright ©2022 by Abdo Consulting Group, Inc. International copyrights reserved in all countries. No part of this book may be reproduced in any form without written permission from the publisher. Cody Koala™ is a trademark and logo of Pop!.

Printed in the United States of America, North Mankato, Minnesota

102021
012022

THIS BOOK CONTAINS RECYCLED MATERIALS

Cover Photo: Shutterstock Images
Interior Photos: Shutterstock Images, 1–8, 11–20; Claude Nuridsany & Marie Perennou/Science Source, 9

Editor: Elizabeth Andrews
Series Designers: Laura Graphenteen, Victoria Bates

Library of Congress Control Number: 2021942277
Publisher's Cataloging-in-Publication Data
Names: Gieseke, Tyler, author.
Title: Turning into a newt / by Tyler Gieseke
Description: Minneapolis, Minnesota : Pop!, 2022 | Series: Transforming animals | Includes online resources and index.
Identifiers: ISBN 9781098241193 (lib. bdg.) | ISBN 9781098241896 (ebook)
Subjects: LCSH: Newts--Juvenile literature. | Amphibians--Juvenile literature. | Animal life cycles--Juvenile literature. | Amphibians--Metamorphosis --Juvenile literature. | Animal Behavior--Juvenile literature.
Classification: DDC 597.8--dc23

Hello! My name is

Cody Koala

Pop open this book and you'll find QR codes like this one, loaded with information, so you can learn even more!

Scan this code* and others like it while you read, or visit the website below to make this book pop.

popbooksonline.com/turn-newt

*Scanning QR codes requires a web-enabled smart device with a QR code reader app and a camera.

Table of Contents

Chapter 1
Transforming Animals 4

Chapter 2
Life as a Larva 10

Chapter 3
Adult Newts 14

Chapter 4
Survival and New Life 18

Making Connections 22
Glossary 23
Index 24
Online Resources 24

Chapter 1

Transforming Animals

A hiker sees a brightly colored newt come out from under a rock. The newt bites a worm. Adult newts like this one live on land. But young newts live in the water.

Newts live in North America, Europe, and parts of Asia.

Watch a video here!

Newts are **transforming** animals. They grow through three steps. The steps are egg, larva, and adult.

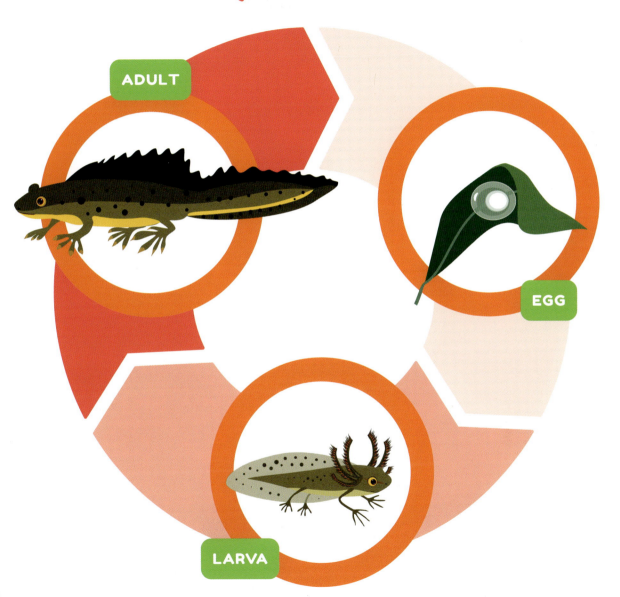

Newts begin life as eggs. Female newts lay their eggs in fresh water, on plants or rocks. The eggs are small and soft.

Chapter 2

Life as a Larva

Newt eggs take two or three weeks to **hatch**. When a newt hatches, it is called a larva. It does not have legs yet. It lives underwater.

Complete an activity here!

The larva eats small insects in the water. It gets stronger and grows legs. **Gills** stick out of its head.

As the larva gets even older, the gills shrink. The larva grows lungs. These will help it breathe air.

Chapter 3

Adult Newts

Newts are **amphibians**. When a larva is a few months old, it leaves the water and lives on land. Now it is an adult newt. Its **gills** are gone.

When newts leave the water, they are usually very small. Some are only as long as a human thumbnail!

Learn more here!

Adult newts stay in cool, moist places, like under plants or rocks. They wait for **prey** to come near.

Then, they strike! Newts eat worms and insects. The newts grow bigger over time.

Chapter 4

Survival and New Life

Snakes and birds eat newts. But newts have brightly colored skin. This tells **predators** the newts are toxic. Then, the predators might leave the newts alone.

Learn more here!

Once a newt is a few years old, it goes back to the water in the spring. There, it finds a **mate**. The two newts make new eggs. The life cycle starts again! Newts are amazing **transforming** animals.

Making Connections

Text-to-Self

Which is your favorite step in the newt life cycle? Why?

Text-to-Text

What other stories or books have you read that include amphibians? How do those texts compare with this one?

Text-to-World

What is another transforming animal you know about? How is its life cycle similar to or different from a newt's?

Glossary

amphibian – an animal that lives both in water and on land.

gills – body parts that help animals breathe water.

hatch – to break out of an egg.

mate – a partner animal of the same kind. Together they make new eggs or babies.

predator – an animal that hunts other animals for food.

prey – an animal that other animals hunt for food.

transform – to change into a new shape.

Index

color, 4, 18

eggs, 6–10, 21

food, 4, 13, 16–17

land, 4, 14

life cycle, 6–7, 21

mate, 21

size, 9, 17

water, 4, 9–10, 13–15, 21

Online Resources

popbooksonline.com

Thanks for reading this Cody Koala book!

Scan this code* and others like it in this book, or visit the website below to make this book pop!

popbooksonline.com/turn-newt

*Scanning QR codes requires a web-enabled smart device with a QR code reader app and a camera.